Work 103

建造大教堂

Building a Cathedral

Gunter Pauli

[比] 冈特·鲍利 著

[哥伦] 凯瑟琳娜·巴赫 绘

姚晨辉 译

上海远东出版社

丛书编委会

主　任：田成川

副主任：何家振　闫世东　林　玉

委　员：李原原　翟致信　靳增江　史国鹏　梁雅丽
　　　　任泽林　陈　卫　薛　梅　王　岢　郑循如
　　　　彭　勇　王梦雨

特别感谢以下热心人士对童书工作的支持：

匡志强　宋小华　解　东　厉　云　李　婧　庞英元
李　阳　刘　丹　冯家宝　熊彩虹　罗淑怡　旷　婉
杨　荣　刘学振　何圣霖　廖清州　谭燕宁　王　征
李　杰　韦小宏　欧　亮　陈强林　陈　果　寿颖慧
罗　佳　傅　俊　白永喆　戴　虹

目录

Contents

在加拉帕

戈斯群岛中的一个小岛

上，孤独的乔治正绕着他的花园散步。他

感到非常寂寞，这时一只大力甲虫走出了岸边的热

带雨林，过来看望他。

大力甲虫问道："这么说，你就一直在同一座

房子里住了100多年？"

Lonesome George is taking a
walk around his garden on one of the
Galápagos Islands. He is feeling very lonely
until a hercules beetle, from the rainforest
on the shore, pays him a visit.

"So, you have been living in the same house
for over one hundred years?" the beetle
asks.

一直在同一座房子里住了100多年

Living in the same house for one hundred years

不幸的是，我是唯一的幸存者了

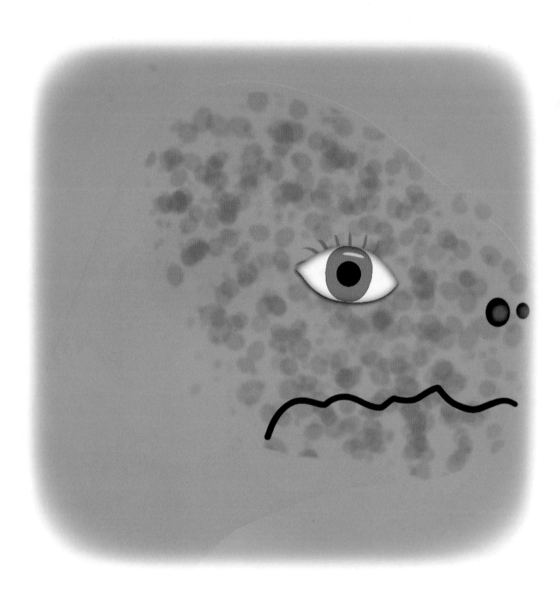

Sadly, I'm the last one left

6

"哦，是啊，这很了不起吧！"

"在一座房子里住100年，真是非常少见。几乎没有谁可以活这么久，也几乎没有哪座房子能这么坚固。"

"的确，要知道，像我这个种类的龟在地球上已经生存了500万年。但不幸的是，我是唯一的幸存者了。"

"Oh yes, and it has been great."

"To live in a house for a hundred years is quite exceptional. Hardly anyone lives that long, and hardly anyone has a house that strong."

"Well, my kind of tortoise has been around for five millions years, you know. Sadly, I'm the last one left."

"你所有的兄弟姐妹和其他家庭成员都去世了吗？他们是生病了吗？"

"不，导致他们死亡的罪魁祸首不是疾病，而是人类！那些捕鲸者和海豹猎手来这里找吃的。因为我们行动缓慢，他们就把我们当作食物。"

"How did all your brothers and sisters and other family members die? Did they get sick?"

"No, it wasn't diseases that killed them, it was people! Those whale hunters and seal hunters, who came here looking for an easy meal. As we cannot move fast, they had us for dinner."

他们就把我们当作食物

They had us for dinner

……将我们活生生地煮了！

... cooking us alive ...

"碰到这样的人，你背上的这座房子不能保护你吗？"

"瞧啊，我们是非常容易就长到100千克重的大家伙。可这并不能阻止一群男人将我们翻转过来并把我们活生生地煮了。我的家人只能任由宰割。"

"Couldn't that house you carry on your back protect you against these people?"

"Look, we can easily weigh up to a hundred kilogram. But that doesn't stop a group of men turning us on our backs and cooking us alive. There was very little my family could do."

"眼睁睁地看着这一切发生肯定是一件非常可怕的事情。"大力甲虫说。

"这还没完。人类还把猫啊老鼠啊带上了岛，这些动物捕杀我们那些无助的孩子。而现在你们这些甲虫也成了捕猎目标！"

"的确是这样，不过没有人喜欢吃我们。他们只是想把我们做成标本挂在墙上。"

"It must have been terrible to see this happening before your eyes," remarks the beetle.

"That's not all. People also brought animals like cats and rats to the islands. They killed off our helpless little ones. And now you beetles are also in demand!"

"That's true, but no one is keen to eat us. They just want to hang us on their walls."

人类还将猫啊老鼠啊带上了岛

People also brought animals like cats and rats
to the islands

我很强壮。

I'm very strong.

"做标本？是因为你的外壳吗？"乌龟问。

"是的，我有一个非常特别的壳。而且我很强壮，我可以举起相当于自身体重500倍的东西。"

"哇，太不可思议了！你曾经为印加人工作过吗？"

"Is it because of your shell?"
the tortoise asks.

"Yes, I have a very special shell. And I'm very strong. I can lift things that are five hundred times my own weight."

"Now that's amazing! Did you ever work for the Incas?"

"印加人？你指的是美洲的那个伟大文明吗？没有，为什么这样问？"

"嗯，他们在山上建造了宏伟的建筑，比如秘鲁的马丘比丘古城，还有哥伦比亚的失落之城。我想你这么强壮，可能会帮他们一把。"

"不，我们生活在这里的岛屿上，住在茂密的沿海森林中。不过，我曾经到那些石头建筑里转了一圈，它们的建造方式让我很惊讶！"

"The Incas? You mean that great culture of the Americas? No, why?"

"Well, they built grand structures in the mountains, such as at Machu Picchu in Peru and the Lost City in Colombia. I thought with your strength you might have given them a hand."

"No, we live here in the thick coastal forest on the islands. I've paid a quick visit to those stone structures though, and was amazed at the way people had built it."

他们在山上建造了宏伟的建筑

They built grand structures in the mountains

成百上千座美丽而宏伟的建筑

Hundreds of beautiful, grand buildings

"我听说，在大西洋对面的欧洲，人们建造了成百上千座美丽而宏伟的建筑，叫作大教堂。有些大教堂花了一个多世纪才建成。这些建筑肯定也值得一看。"

"是的，当时的欧洲人有着坚定的信念，那些开始建设的人很清楚，直到死去他们也不一定能亲眼看到建筑完工。他们几代人辛勤工作，最终将他们的宏伟梦想变成现实，并请他们那个时代最伟大的艺术家装饰教堂内部。"

"I heard that across the Atlantic Ocean, in Europe, people have built hundreds of beautiful, grand buildings, called cathedrals. Some took over a century to complete. Now those must be worth seeing as well."

"Yes, the people of Europe had so much faith that those who started building knew they would not see the end result in their lifetime. They dedicated generations of work to make their grand dreams come true and used some of the greatest artists of their time to decorate the interior."

"有这种信念一定很棒。我多么希望能去看一看那些建筑，并与家人分享我看到的美景。但这个愿望恐怕不可能实现了。"孤独的乔治叹息道。

"虽然我只是一只甲虫，但如果你愿意的话，我会陪你一起去的。"大力甲虫提议。

……这仅仅是开始！……

"It must be good to have faith like that. I wish I could see one of those buildings, and share the sight with someone from my family. But I'm afraid that's no longer possible," sighs Lonesome George.

"I know I am just a beetle, but I'll come with you if you want," offers Hercules.

... AND IT HAS ONLY JUST BEGUN!...

·······这仅仅是开始！·······

... AND IT HAS ONLY JUST BEGUN! ...

建成一座大教堂平均要花一个多世纪的时间。天主教会拥有5000多名主教和大约12亿名教徒。主教们在大教堂有自己的座位。

The construction of a cathedral took, on average, more than a century. The Catholic Church has more than five thousand bishops for the 12 billion followers of the Catholic faith. Bishops have their seats in cathedrals.

The Cathedral of Chartres (France) took only 25 years to build. The Sagrada Familia (Sacred Family) Cathedral in Barcelona has been under construction for more than a hundred years and has been funded by donations only.

法国的沙特尔大教堂只用了25年就建成了。巴塞罗那的圣家族大教堂已经动工100多年，现在仍在建造当中，资金来源仅仅依靠捐款。

孤独的乔治是一只雄性平塔岛象龟，在102岁的时候死亡。它是该物种的最后一名成员，被安置在厄瓜多尔加拉帕戈斯群岛的查尔斯·达尔文研究站中。乔治去世后，平塔岛象龟被宣布灭绝。

Lonesome George, a male Pinta Island tortoise, was 102 years old when he died. He was the last member of his species and was kept at the Charles Darwin Research Station on the Galápagos Islands (Ecuador). When George died, the Pinta Island tortoise was declared extinct.

102 岁

巨型陆龟是从大陆来到岛屿上的。它们依靠浮力，游水的时候能将脖子伸出水面进行呼吸，再加上在没有食物或淡水时仍能存活数月的本领，成功地长途跋涉1000千米。

Giant tortoises came to the islands from the mainland. Aided by their buoyancy, their ability to breathe while swimming by extending their necks above water, and their strength to survive for months without food or fresh water, they were able to survive the 1,000 km long journey.

大力甲虫的幼虫期长达一年甚至更长，在此期间主要吃腐烂的木头。成年之后，大力甲虫在南美洲和中美洲森林的地面上游荡，靠吃腐烂的水果为生。

The hercules beetle spends a year or more of its life as a larva, eating rotting wood. As an adult, the beetle roams the forest floor in South America and Central America, eating rotting fruit.

大力甲虫可以举起850倍于自身重量的物体。如果人类有这样的本领，一个人就能够举起65吨重的物体。不过，蜣螂（俗称屎壳郎）更为强壮，它能够拖动1 100倍于自身体重的物体。

The hercules beetle can lift an object 850 times its own weight. If a human had that strength, a man would be able to lift 65 tonnes. The dung beetle is even stronger; it is able to pull 1,100 times its own body weight.

马丘比丘（秘鲁）在克丘亚语中是"老年人"的意思，该城建造于15世纪。马丘比丘被认为是世界七大奇迹之一，并在1983年被联合国教科文组织宣布为世界遗产。

Machu Picchu (Peru), which in the Quechua language means "old person", was built in the 15th century. It is considered one of the Seven Wonders of the World and was declared a UNESCO World Heritage site in 1983.

圣玛尔塔失落之城（哥伦比亚）的建造比马丘比丘还要早650年，它由雕凿于山坡上的一系列台阶组成，并由铺设的道路网络相连接。

The Lost City of Santa Martha (Colombia) was built 650 years earlier than Machu Picchu. It consists of a series of terraces carved into the mountainside, with a network of tiled roads.

如果你知道自己永远不会看到某件事情的最终结果，你还会下决心为其投入资源吗？如果连你的孩子也没有机会看到你的远见卓识和奉献精神的成果，那又会怎样？

Could you make the decision to dedicate resources to something of which you know you'll never see the end result? And what if not even your children will see the fruits of your vision and dedication?

你是否曾经有机会吃一只大乌龟的肉？它最终幸免于难了吗？

Would you have liked to eat the meat of a big turtle, had they survived?

Do you think that hanging a dead hercules beetle on the wall will make your house look beautiful?

你觉得在墙上挂一只死去的大力甲虫会让房间看上去更漂亮吗？

整个一生都将自己的房子背在背上会是一件很享受的事情吗？或者说，你会认为这是一种挑战吗？

Is it a luxury to carry your own house on your back for your entire life? Or would you consider it a challenge?

Do It Yourself!

自己动手！

Lost cities like Santa Martha and Machu Picchu were built of rock. How many houses and buildings around your home are built of rock? Is anyone in your area building a new house using rocks and stone? Find out why construction with rocks and stone has lost favour and why people lost interest in using this as a building material. Stone buildings last for hundreds of years, which makes them very economical as the capital costs are spread over a few generations. Now ask around to find out if anyone would be prepared to build a house that will last for generations. What do you think the answer will be?

像圣玛尔塔和马丘比丘这样消失的城市是用岩石建成的。你家的周围有多少房子或建筑是用岩石建造的？你所在的地区有没有人用岩石和石头建造新房子？找出原因，为什么用岩石和石头建造的建筑已经不受欢迎了，人们为什么对石头这种建筑材料失去了兴趣。石头建筑能保存几百年，它们的建设成本可以分摊到几代人身上，因此非常经济。问一问周围的人，是否有人准备盖一座可以延续几代的房子。你认为会得到什么答案？

学科知识

Academic Knowledge

生物学	海龟和陆龟都属于龟鳖目；变温动物，或称冷血动物，指动物的体温随外界温度变化而改变；外骨骼是指脊柱和肋骨融合成为一个外壳；乌龟的外壳是由骨骼组成的，最外面的保护层是由角质组成的；大力甲虫是夜行性的，目的是躲避天敌。
化 学	角质在构建乌龟龟甲保护层中的重要性。
物 理	异速生长，有关身体大小和形状、解剖学、生理学和行为之间关系的研究；大力甲虫的外壳在干燥时为绿色，潮湿时会变成黑色，利用薄膜干涉效应增加湿度；大力甲虫的飞行能力并没有受到组成其身体的轻质材料的影响；尖锐摩擦声是指昆虫摩擦其身体部位而形成的吱吱的噪声。
工程学	建筑技术的演变决定了教堂不同的建筑风格，包括罗马式、哥特式、新哥特式和巴洛克式风格。
经济学	对教堂的投资往往没有经济回报，而是对艺术、建筑学和建筑的支持。
伦理学	由人类造成的物种灭绝大潮；为某一目标而奋斗的承诺，在你的有生之年可能都无法亲眼看到该目标的实现或享受目标的成果；虽然我们都知道物种的灭绝是由人类的行为造成的，但我们似乎并不准备为此作出必要的改变；很多人认为动物是在为人类服务，我们有权利去杀死甚至灭绝它们。
历 史	大教堂在公元313年首次成为主教座堂；在希腊神话中，赫拉克勒斯是最强壮的凡人，是宙斯最后一个凡人儿子，他曾经花费大量时间为他的愤怒和鲁莽行为进行忏悔；印加文明。
地 理	加拉帕戈斯群岛；马丘比丘位于秘鲁；圣玛尔塔失落之城位于哥伦比亚；大西洋。
数 学	两个量之间的函数关系是指其中一个量的变化会导致另一个量的相应变化。
生活方式	我们目前的生产和消费体系会导致某些物种的灭绝，需要对生活方式作出改变，以避免这种事情的发生；高领衫或高领毛衣是受到乌龟形态的启发。
社会学	种族灭绝是指针对一个民族的毁灭意图和行为或大屠杀，但我们并不将对动物进行大规模杀害而导致其灭绝的行为称为种族灭绝。
心理学	信仰的力量，有了信念的支持，人们就能够坚持不懈，排除万难。
系统论	人类往往不清楚自己的行为会带来的意外后果，如从人类的食物生产方式到非本地物种引进等。

情感智慧
Emotional Intelligence

孤独的乔治

当乔治谈到住在自己的房子里很棒时，似乎对生活感到很满意，但他知道他已经是自己种族的最后一个成员，并敏锐地意识到种族即将灭绝的困境。更糟糕的是，乔治亲眼目睹了人类毁灭他的种族的"暴行"。乔治接受了自己的局限性，承认自己没有办法去保护其他同类。乔治在思考人类给他们的种族带来的混乱状况及其日益增长的复杂性。他很清楚大力甲虫的起源和能力，知道他们也在被人类捕杀，并对大力甲虫表示了同情。乔治思考着对宏伟建筑的建设，感叹着梦想的力量以及愚公移山的献身精神。乔治冷静地接受了这个事实：作为种族的最后一个成员，他将无法再和同类分享任何东西。

大力甲虫

甲虫同情乔治，认识到了他的独特性。甲虫急于了解乔治的困境，震惊于坚硬的龟壳居然不能保护那些乌龟免遭猎人的杀害。他很遗憾已经发生的悲剧，更为乔治的现状感到无奈。乔治关于甲虫和古老文明之间可能存在联系的哲学思考，给甲虫留下了深刻印象。了解到那些教堂建造者基于自己的信念，几代人坚持不懈地建造教堂，甲虫表达了深深的敬意。甲虫也对孤独的乔治的感情表示了尊重，并提出可以陪他去参观大教堂。

艺术
The Arts

当我们观看大力甲虫及其近亲独角仙的照片时，我们可以看到这些物种别具一格的形状和轮廓的神奇之处。它们不仅具有有趣的形状，还有两套翅膀，甚至可以改变颜色。这一点可以启发我们重新创造这些神奇生物的形状和色调。找五六张不同甲虫的图片，仔细研究。然后盖住所有的图片，凭借记忆用钢笔或铅笔把它们画出来，重新创造每个甲虫的形状和轮廓。你的作品会比真正的甲虫更吓人吗？

思维拓展
Systems: Making the Connections

我们一般希望建筑物能有较长的使用寿命。石头和岩石的建筑物建设周期很长，往往需要几十年甚至上百年，但也会保存几千年时间。在世界各地，人类建造了一些体现他们信仰和文化的建筑物。这些建设大部分都基于一种信念，即拥有这样的建筑对社会是有利的。在这种信念的支持下，财力和人力资源被调动起来，这些长期决策获得了跨时代的支持，最终使得后人可以享受成功的果实。实施可以服务社会的长期项目的能力是我们不应该失去的，比如建造大教堂这样在几百年后仍可为社会作贡献的工程。与此同时，人们似乎已经失去了引导社会可持续发展的意识。有几个因素造成了众多物种的大规模消失：一是为满足眼前的营养需求而杀害物种（如杀死乌龟做食物）；二是引进非本地物种对当地生物多样性造成严重破坏；此外还有用动物"战利品"装饰房间的欲望，对于这一点，人们看重的是动物生命价值的抽象意义。第一种情况是即时需要，第二种是相当无知的行为，最后一种情况只是为了取悦自我。人类的这种破坏性力量与他们创造文化、建造文物、设置长期目标并历经几代人去实现的独特能力形成了巨大的反差。人们需要重建他们作为地球护卫者的信念，并确保每一个物种都有生存的空间，因为地球上每一个物种的积极影响将在几代人之后呈现出来。

动手能力
Capacity to Implement

设想一个你一生可能都无法达到的目标或完成的事情。你现在可以做些什么，以确保未来的三代人为实现你确定的目标而继续努力？

故事灵感来自
This Fable Is Inspired by

玛丽·拉萨特
Marie Rassart

玛丽·拉萨特 2010 年毕业于那慕尔大学，获得了物理学博士学位。她专门从事光学研究，分析复杂的纳米结构。她曾与让·珀尔·维涅龙（Jean Pol Vigneron）教授共事，维涅龙教授研究过高山花卉用什么方式散射而不是拦截紫外线（这是本童书第100个童话《阳光下的雪绒花》的灵感来源）。玛丽学到了如何成功地复制具有相同光学性能的天然多层纳米结构。她是一名年轻的研究人员，愿意通过科学教育和终身学习，利用从大自然中获得的灵感，帮助企业在盈利的同时实现可持续发展。

图书在版编目（CIP）数据

冈特生态童书.第三辑修订版：全36册：汉英对照 /
（比）冈特·鲍利著；（哥伦）凯瑟琳娜·巴赫绘；
何家振等译.—上海：上海远东出版社，2022
书名原文：Gunter's Fables
ISBN 978-7-5476-1850-9

Ⅰ.①冈… Ⅱ.①冈…②凯…③何… Ⅲ.①生态环
境–环境保护–儿童读物—汉、英 Ⅳ.①X171.1-49

中国版本图书馆CIP数据核字（2022）第163904号
著作权合同登记号图字09-2022-0637号

策　　划 张　蓉
责任编辑 程云琦
封面设计 魏　来 李　廉

冈特生态童书

建造大教堂

[比]冈特·鲍利　著
[哥伦]凯瑟琳娜·巴赫　绘

姚晨辉　译

记得要和身边的小朋友分享环保知识哦！
八喜冰淇淋祝你成为环保小使者！